Let's Add!

MW01041844

Find the sum.

1
5 + 3 = 8
50 + 30 = ■

2
4 + 3 = 7
400 + 300 = ■

3
2 + 6 = 8
200 + 600 = ■

4
1 + 8 = 9
10 + 80 = ■

5
8 + 6 = 14
80 + 60 = ■

6
5 + 7 = 12
500 + 700 = ■

7
5 + 4 = 9
500 + 400 = ■

8
7 + 7 = 14
700 + 700 = ■

9
4 + 8 = 12
40 + 80 = ■

10
8 + 7 = 15
80 + 70 = ■

11
2 + 5 = 7
20 + 50 = ■

12
9 + 6 = 15
900 + 600 = ■

Answer Box

A	B	C	D	E	F
1,200	1,400	800	70	90	150
G	**H**	**I**	**J**	**K**	**L**
1,500	700	140	120	900	80

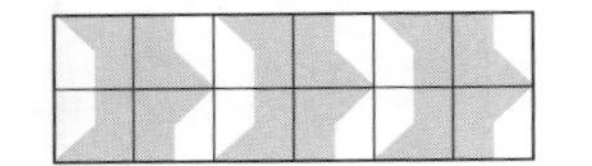

Objective: Use a basic fact to find the sum of tens or hundreds.

Addition Fun!

Example

Find the sum.

42 + 6

Think: **2 + 6 = 8**, so **42 + 6 = 48.**

Find the sum.

1 33 + 6

2 41 + 7

3 25 + 4

4 72 + 5

5 63 + 2

6 23 + 4

7 42 + 3

8 77 + 2

9 31 + 6

10 64 + 4

11 When 3 is added to 46, what is the sum?

12 What is the sum of ninety-four and five?

Answer Box

A 45	B 68	C 65	D 37	E 99	F 79
G 77	H 39	I 29	J 49	K 27	L 48

Objective: Find the sum of a 2-digit number and a 1-digit number without regrouping.

Adding Ones, Then Tens

Example

Find the sum.

$$\begin{array}{r} 32 \\ +\ 41 \\ \hline 3 \end{array}$$ Add the ones.

$$\begin{array}{r} 32 \\ +\ 41 \\ \hline 73 \end{array}$$ Then add the tens.

Find the sum.

1 $\begin{array}{r} 45 \\ +\ 31 \\ \hline \end{array}$

2 $\begin{array}{r} 24 \\ +\ 61 \\ \hline \end{array}$

3 $\begin{array}{r} 51 \\ +\ 48 \\ \hline \end{array}$

4 $\begin{array}{r} 36 \\ +\ 41 \\ \hline \end{array}$

5 What is the sum of seventy-six and twenty-one?

6 When 23 is added to 25, what is the sum?

7 $\begin{array}{r} 63 \\ +\ 33 \\ \hline \end{array}$

8 $\begin{array}{r} 40 \\ +\ 47 \\ \hline \end{array}$

9 $\begin{array}{r} 36 \\ +\ 13 \\ \hline \end{array}$

10 $\begin{array}{r} 35 \\ +\ 22 \\ \hline \end{array}$

11 What is the sum of fifty-two and twenty-six?

12 What is the sum of forty-four and eleven?

Answer Box

A 49	B 99	C 76	D 77	E 85	F 48
G 97	H 78	I 96	J 57	K 55	L 87

Objective: Find the sum of two 2-digit numbers without regrouping.

Wow! Adding Hundreds!

First add ones.
Then add tens.
Then add hundreds.

Find the sum.

1
$$\begin{array}{r} 342 \\ +\ 406 \\ \hline \end{array}$$

2
$$\begin{array}{r} 536 \\ +\ 223 \\ \hline \end{array}$$

3
$$\begin{array}{r} 487 \\ +\ 102 \\ \hline \end{array}$$

4
$$\begin{array}{r} 156 \\ +\ 521 \\ \hline \end{array}$$

5
$$\begin{array}{r} 543 \\ +\ 355 \\ \hline \end{array}$$

6
$$\begin{array}{r} 326 \\ +\ 451 \\ \hline \end{array}$$

7
$$\begin{array}{r} 430 \\ +\ 369 \\ \hline \end{array}$$

8
$$\begin{array}{r} 505 \\ +\ 381 \\ \hline \end{array}$$

9 231 and 147 are added. What is the sum?

10 What is the sum of 326 and 302?

11 What is the sum of 639 and 240?

12 176 and 212 are added. What is the sum?

Answer Box

A	B	C	D	E	F
777	759	677	748	589	378
G	**H**	**I**	**J**	**K**	**L**
388	886	799	628	898	879

Objective: Find the sum of two 3-digit numbers without regrouping.

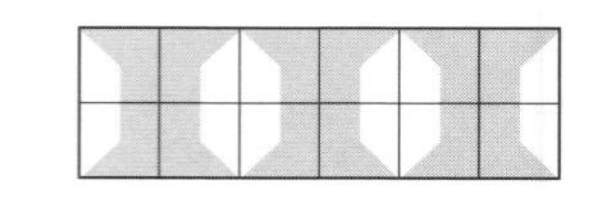

Round Up

Round to the nearest ten.

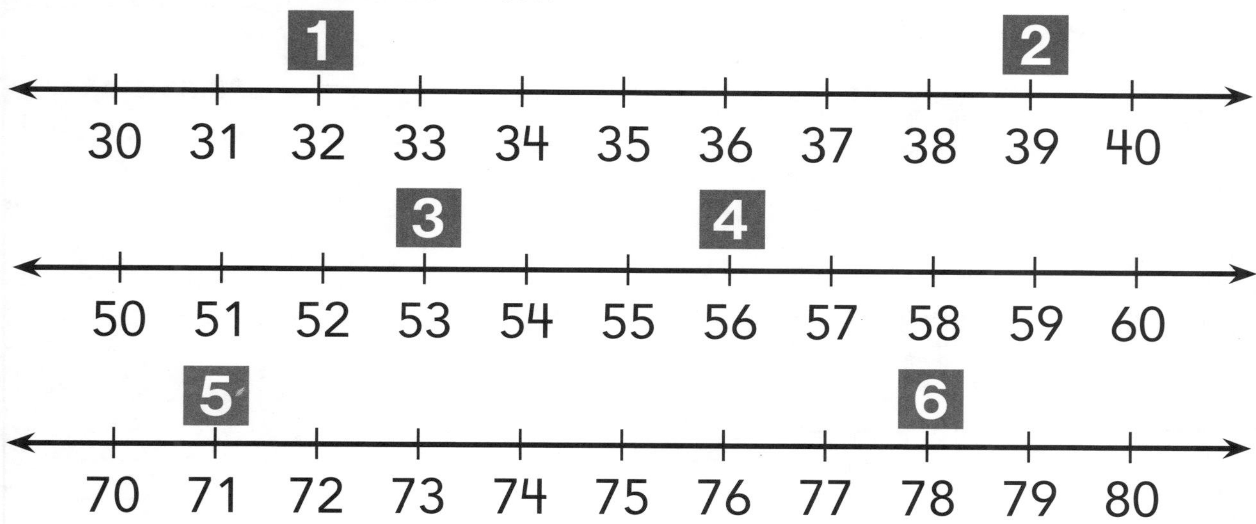

Round to the nearest hundred.

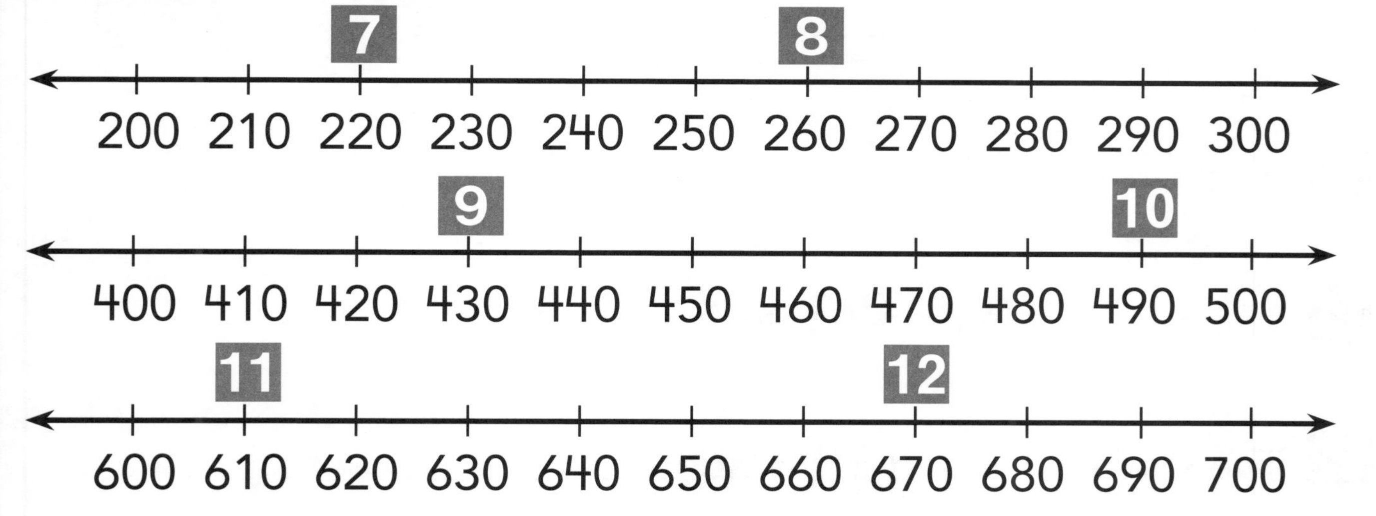

A 200	B 500	C 700	D 300	E 70	F 600
G 60	H 30	I 80	J 40	K 50	L 400

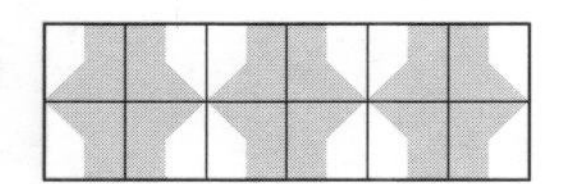

Objective: Round a 2- or 3-digit number to the nearest ten or hundred.

Close Enough!

Round each addend to the nearest ten.
Then find the best estimate.

You can estimate by adding rounded numbers!

1 $\begin{array}{r} 34 \\ +\ 48 \\ \hline \end{array}$

2 $\begin{array}{r} 26 \\ +\ 57 \\ \hline \end{array}$

3 $\begin{array}{r} 12 \\ +\ 22 \\ \hline \end{array}$

4 $\begin{array}{r} 62 \\ +\ 51 \\ \hline \end{array}$

5 $\begin{array}{r} 49 \\ +\ 12 \\ \hline \end{array}$

6 $\begin{array}{r} 21 \\ +\ 32 \\ \hline \end{array}$

7 $\begin{array}{r} 23 \\ +\ 19 \\ \hline \end{array}$

8 $\begin{array}{r} 64 \\ +\ 44 \\ \hline \end{array}$

9 $\begin{array}{r} 33 \\ +\ 39 \\ \hline \end{array}$

10 Add 63 and 61. The sum is about ▢.

11 The sum of 63 and 72 is about ▢.

12 The sum of 11 and 14 is about ▢.

Answer Box

A 60	B 70	C 20	D 90	E 40	F 80
G 30	H 130	I 50	J 100	K 110	L 120

Objective: Find the estimated sum of two 2-digit numbers.

A New Group

Find the missing number.

1. 5 tens 14 ones = 6 tens ▢ ones
2. 8 tens 14 ones = ▢ tens 4 ones
3. 0 tens 19 ones = ▢ ten 9 ones
4. 1 ten 0 ones = 0 tens ▢ ones
5. 7 tens 11 ones = ▢ tens 1 one
6. 6 tens 17 ones = ▢ tens 7 ones
7. 13 tens 18 ones = ▢ tens 8 ones
8. 2 tens 12 ones = 3 tens ▢ ones
9. 2 tens 14 ones = ▢ tens 4 ones
10. 10 tens 15 ones = ▢ tens 5 ones
11. 7 tens 16 ones = 8 tens ▢ ones
12. 4 tens 16 ones = ▢ tens 6 ones

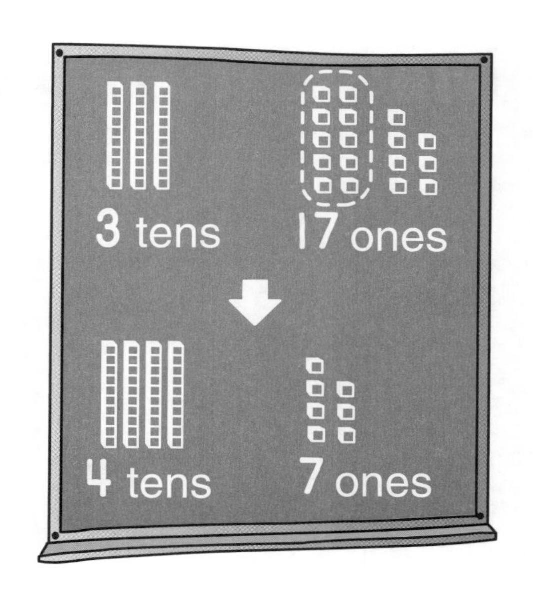

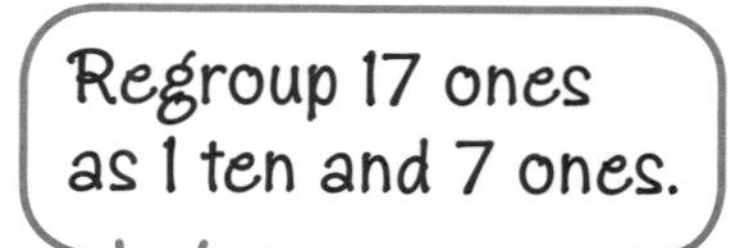

Answer Box

A	B	C	D	E	F
2	5	1	3	11	14
G	**H**	**I**	**J**	**K**	**L**
7	9	6	8	10	4

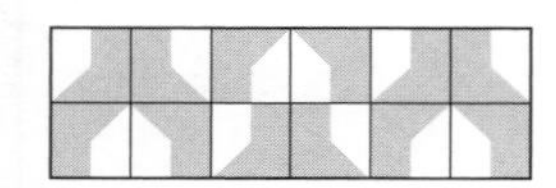

Objective: Regroup a 2-digit number as tens and ones.

Now Add the Groups!

Find the sum.

1
$$\begin{array}{r} 43 \\ +\ 48 \\ \hline \end{array}$$

2
$$\begin{array}{r} 54 \\ +\ 26 \\ \hline \end{array}$$

3
$$\begin{array}{r} 15 \\ +\ 37 \\ \hline \end{array}$$

4
$$\begin{array}{r} 29 \\ +\ 25 \\ \hline \end{array}$$

5
$$\begin{array}{r} 34 \\ +\ 16 \\ \hline \end{array}$$

6
$$\begin{array}{r} 22 \\ +\ 49 \\ \hline \end{array}$$

7
$$\begin{array}{r} 47 \\ +\ 49 \\ \hline \end{array}$$

8
$$\begin{array}{r} 32 \\ +\ 28 \\ \hline \end{array}$$

9 18 and 63 are added. What is their sum?

10 Twenty-six and eighteen are added. What is their sum?

11 28 and 58 are added. What is their sum?

12 What is the sum of thirty-six and twenty-eight?

Answer Box

A 54	B 91	C 50	D 86	E 71	F 80
G 96	H 44	I 81	J 52	K 64	L 60

Objective: Find the sum of two 2-digit numbers with regrouping.

Keep on Adding!

Find the sum.

1
$$\begin{array}{r} 24 \\ +\ 18 \\ \hline \end{array}$$

2
$$\begin{array}{r} 56 \\ +\ 26 \\ \hline \end{array}$$

3
$$\begin{array}{r} 33 \\ +\ 58 \\ \hline \end{array}$$

4
$$\begin{array}{r} 49 \\ +\ 37 \\ \hline \end{array}$$

5
$$\begin{array}{r} 26 \\ +\ 28 \\ \hline \end{array}$$

6
$$\begin{array}{r} 49 \\ +\ 27 \\ \hline \end{array}$$

7
$$\begin{array}{r} 53 \\ +\ 18 \\ \hline \end{array}$$

8
$$\begin{array}{r} 26 \\ +\ 59 \\ \hline \end{array}$$

9
$$\begin{array}{r} 44 \\ +\ 16 \\ \hline \end{array}$$

10 16 plus 16

11 22 plus 48

12 28 plus 16

Answer Box

A	B	C	D	E	F
70	54	32	71	85	44
G	**H**	**I**	**J**	**K**	**L**
91	60	86	42	76	82

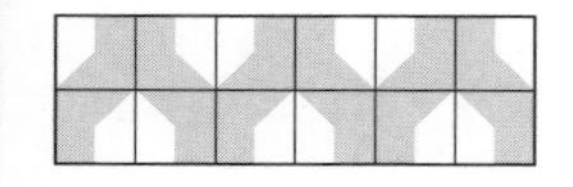

Objective: Find the sum of two 2-digit numbers with regrouping.

Sum Estimating!

Round each addend to the nearest hundred. Then add to find the best estimate.

1. $336 + 102$
2. $660 + 590$
3. $392 + 780$
4. $436 + 275$
5. $859 + 541$
6. $644 + 202$
7. $721 + 328$
8. $537 + 107$
9. $110 + 137$
10. $545 + 611$
11. $507 + 429$
12. $323 + 185$

Answer Box

A	B	C	D	E	F
700	400	500	1,300	1,200	200
G	**H**	**I**	**J**	**K**	**L**
1,100	1,000	800	600	900	1,400

Objective: Find the estimated sum of two 3-digit numbers.

Another Group Is Here!

Find the missing number.

1. 6 hundreds 18 tens = 7 hundreds ▢ tens
2. 10 hundreds 12 tens = ▢ hundreds 2 tens
3. 3 hundreds 16 tens = 4 hundreds ▢ tens
4. 2 hundreds 11 tens = ▢ hundreds 1 ten
5. 4 hundreds 19 tens = 5 hundreds ▢ tens
6. 11 hundreds 14 tens = ▢ hundreds 4 tens
7. 3 hundreds 18 tens = ▢ hundreds 8 tens
8. 1 hundred 0 tens = 0 hundreds ▢ tens
9. 5 hundreds 27 tens = 6 hundreds ▢ tens
10. 4 hundreds 11 tens = 5 hundreds ▢ ten
11. 7 hundreds 32 tens = 8 hundreds ▢ tens
12. 4 hundreds 16 tens = ▢ hundreds 6 tens

Answer Box

A	B	C	D	E	F
9	22	11	12	4	1
G	**H**	**I**	**J**	**K**	**L**
17	6	10	5	8	3

Objective: Regroup a 3-digit number as hundreds and tens.

Some More Sums!

Find the sum.

1 $259 + 319

2 $157 + 128

3 $706 + 244

4 $673 + 118

5 $536 + 408

6 $423 + 139

7 $106 + 467

8 $317 + 623

9 $536 + 245

10 $215 + 626

11 $325 + 547

12 $427 + 507

Answer Box

A	B	C	D	E	F
$285	$562	$872	$944	$578	$791
G	**H**	**I**	**J**	**K**	**L**
$934	$940	$950	$781	$573	$841

Objective: Find the sum of two 3-digit money amounts with regrouping ones as tens.

Time to Regroup

Sometimes you need to regroup tens as hundreds.

Find the sum.

1 $\begin{array}{r} 248 \\ +\ 371 \\ \hline \end{array}$

2 $\begin{array}{r} 450 \\ +\ 364 \\ \hline \end{array}$

3 $\begin{array}{r} 726 \\ +\ 193 \\ \hline \end{array}$

4 $\begin{array}{r} 643 \\ +\ 285 \\ \hline \end{array}$

5 $\begin{array}{r} 138 \\ +\ 170 \\ \hline \end{array}$

6 $\begin{array}{r} 394 \\ +\ 282 \\ \hline \end{array}$

7 $\begin{array}{r} 381 \\ +\ 194 \\ \hline \end{array}$

8 $\begin{array}{r} 257 \\ +\ 562 \\ \hline \end{array}$

9 $\begin{array}{r} 117 \\ +\ 691 \\ \hline \end{array}$

10 $\begin{array}{r} 687 \\ +\ 142 \\ \hline \end{array}$

11 $\begin{array}{r} 276 \\ +\ 242 \\ \hline \end{array}$

12 $\begin{array}{r} 593 \\ +\ 121 \\ \hline \end{array}$

Answer Box

A 829	B 928	C 808	D 308	E 819	F 676
G 619	H 575	I 518	J 919	K 814	L 714

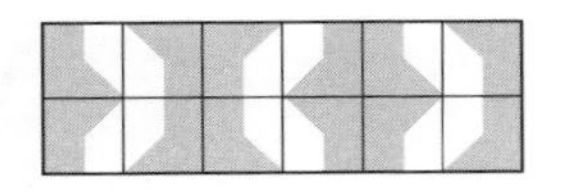

Objective: Find the sum of two 3-digit numbers with regrouping tens as hundreds.

Solve with Addition

Solve.

1 The students sold 241 tickets on Friday.
They sold 245 tickets on Saturday.
How many tickets were sold on both days?

2 390 people buy their tickets early.
147 people buy their tickets at the door.
How many people buy tickets in all?

3 The school prints 230 blue programs.
It prints 250 red programs.
How many programs does the school print in all?

4 Class 2A sells 382 bags of popcorn.
Class 2B sells 326 bags.
How many bags of popcorn do they sell in all?

5 There are 15 green hats and 19 yellow hats.
How many hats are there in all?

6 There are 24 girls and 28 boys.
How many children are there in all?

7 There are 17 boys and 25 girls dancing in the play.

How many children dance in the play?

8 Alice practices her part in the play 6 hours each week.

How much does she practice in 2 weeks?

9 Carlos and Keiko travel 52 miles one way to see the play.

How many miles do they travel both ways?

10 The children in the play practice for 30 minutes each day.

How many minutes do they practice in 2 days?

11 The children sell 138 small T-shirts.

They sell 142 large T-shirts.

How many T-shirts do they sell in all?

12 The school sells pictures of the play.

It sells 292 pictures in May and 136 pictures in June.

How many pictures does the school sell in all?

Answer Box

A 537	B 52	C 708	D 486	E 104	F 480
G 12	H 428	I 42	J 60	K 280	L 34

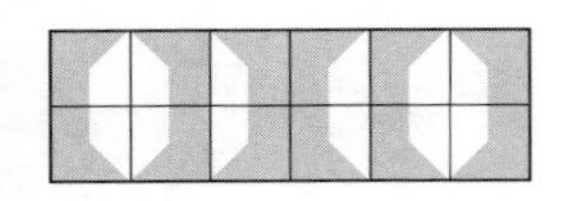

Objective: Solve a problem that involves addition.

Back to Basic Facts!

Find the difference.

1
8 − 2 = 6
80 − 20 = ■

2
9 − 7 = 2
900 − 700 = ■

3
8 − 1 = 7
80 − 10 = ■

4
5 − 2 = 3
500 − 200 = ■

5
17 − 9 = 8
170 − 90 = ■

6
12 − 5 = 7
1,200 − 500 = ■

7
6 − 2 = 4
600 − 200 = ■

8
6 − 3 = 3
60 − 30 = ■

9
16 − 8 = 8
1,600 − 800 = ■

10
9 − 3 = 6
900 − 300 = ■

11
11 − 6 = 5
1,100 − 600 = ■

12
8 − 6 = 2
80 − 60 = ■

Answer Box

A	B	C	D	E	F
30	20	600	400	500	80
G	**H**	**I**	**J**	**K**	**L**
200	700	60	300	800	70

Objective: Use a basic fact to find the difference of tens or hundreds.

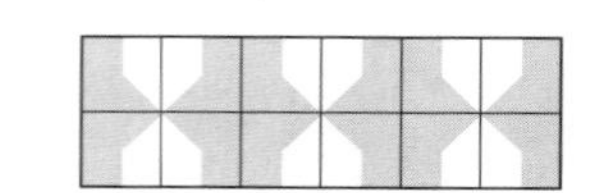

Let's Subtract!

Example

Find the difference.

38 – 2

Think: **8 – 2 = 6**, so **38 – 2 = 36.**

Use a basic fact to help you subtract.

Find the difference.

1 59 – 6

2 75 – 2

3 49 – 8

4 67 – 5

5 53 – 2

6 88 – 8

7 78 – 3

8 64 – 3

9 85 – 4

10 48 – 5

11 95 – 4

12 99 – 6

Answer Box

A	B	C	D	E	F
62	43	73	80	51	81
G	**H**	**I**	**J**	**K**	**L**
75	53	93	61	41	91

Objective: Find the difference of a 2-digit number and a 1-digit number without regrouping.

Get Ready for Tens!

Example

Find the difference.

$$\begin{array}{r} 48 \\ -\ 25 \\ \hline 3 \end{array}$$ Subtract the ones.

$$\begin{array}{r} 48 \\ -\ 25 \\ \hline 23 \end{array}$$ Then subtract the tens.

Find the difference.

1 $\begin{array}{r} 59 \\ -\ 27 \\ \hline \end{array}$

2 $\begin{array}{r} 88 \\ -\ 37 \\ \hline \end{array}$

3 $\begin{array}{r} 74 \\ -\ 51 \\ \hline \end{array}$

4 $\begin{array}{r} 37 \\ -\ 15 \\ \hline \end{array}$

5 $\begin{array}{r} 43 \\ -\ 22 \\ \hline \end{array}$

6 $\begin{array}{r} 29 \\ -\ 18 \\ \hline \end{array}$

7 $\begin{array}{r} 56 \\ -\ 23 \\ \hline \end{array}$

8 $\begin{array}{r} 90 \\ -\ 40 \\ \hline \end{array}$

9 96 take away 30 is ■.

10 86 take away 74 is ■.

11 Ninety-eight take away thirty-six is ■.

12 Seventy-five take away twelve is ■.

Answer Box

A	B	C	D	E	F
21	66	63	50	22	12
G	**H**	**I**	**J**	**K**	**L**
23	62	11	51	33	32

Objective: Find the difference of two 2-digit numbers without regrouping.

Presenting the Hundreds!

First subtract ones.
Then subtract tens.
Then subtract hundreds.

Find the difference.

1
$$\begin{array}{r} 547 \\ -\ 124 \\ \hline \end{array}$$

2
$$\begin{array}{r} 848 \\ -\ 347 \\ \hline \end{array}$$

3
$$\begin{array}{r} 795 \\ -\ 420 \\ \hline \end{array}$$

4
$$\begin{array}{r} 636 \\ -\ 225 \\ \hline \end{array}$$

5
$$\begin{array}{r} 968 \\ -\ 637 \\ \hline \end{array}$$

6
$$\begin{array}{r} 480 \\ -\ 210 \\ \hline \end{array}$$

7
$$\begin{array}{r} 758 \\ -\ 341 \\ \hline \end{array}$$

8
$$\begin{array}{r} 527 \\ -\ 205 \\ \hline \end{array}$$

9
$$\begin{array}{r} 869 \\ -\ 468 \\ \hline \end{array}$$

10
$$\begin{array}{r} 246 \\ -\ 132 \\ \hline \end{array}$$

11
$$\begin{array}{r} 841 \\ -\ 540 \\ \hline \end{array}$$

12
$$\begin{array}{r} 765 \\ -\ 442 \\ \hline \end{array}$$

Answer Box

A	B	C	D	E	F
501	270	401	375	423	411
G	**H**	**I**	**J**	**K**	**L**
322	323	301	331	114	417

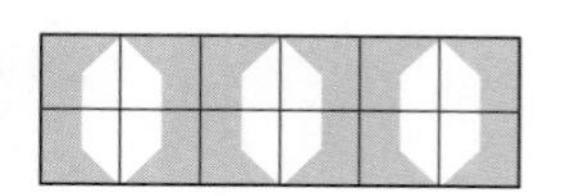

Objective Find the difference of two 3-digit numbers without regrouping.

What's the Difference?

Round each number to the nearest ten. Then find the best estimate.

1
$$\begin{array}{r} 63 \\ -\ 19 \\ \hline \end{array}$$

2
$$\begin{array}{r} 89 \\ -\ 33 \\ \hline \end{array}$$

3
$$\begin{array}{r} 92 \\ -\ 41 \\ \hline \end{array}$$

4
$$\begin{array}{r} 97 \\ -\ 11 \\ \hline \end{array}$$

5
$$\begin{array}{r} 123 \\ -\ 13 \\ \hline \end{array}$$

6
$$\begin{array}{r} 87 \\ -\ 71 \\ \hline \end{array}$$

7
$$\begin{array}{r} 98 \\ -\ 19 \\ \hline \end{array}$$

8
$$\begin{array}{r} 53 \\ -\ 22 \\ \hline \end{array}$$

9
$$\begin{array}{r} 122 \\ -\ 19 \\ \hline \end{array}$$

10
$$\begin{array}{r} 74 \\ -\ 61 \\ \hline \end{array}$$

11
$$\begin{array}{r} 88 \\ -\ 23 \\ \hline \end{array}$$

12
$$\begin{array}{r} 145 \\ -\ 27 \\ \hline \end{array}$$

Answer Box

A	B	C	D	E	F
10	90	70	110	60	120
G	**H**	**I**	**J**	**K**	**L**
40	80	50	100	30	20

Objective: Find the estimated difference of two 2-digit numbers, or a 2- and 3-digit number.

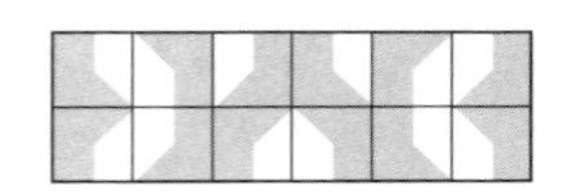

It's Still the Same!

Find the missing number.

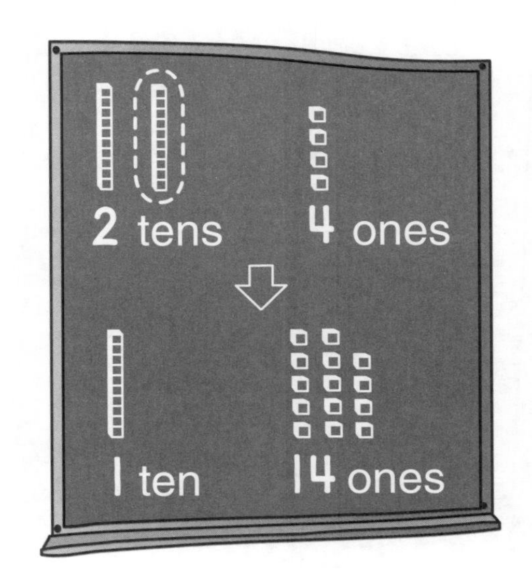

1. 3 tens 5 ones = 2 tens ■ ones
2. 7 tens 7 ones = 6 tens ■ ones
3. 4 tens 0 ones = ■ tens 10 ones
4. 6 tens 8 ones = ■ tens 18 ones
5. 3 tens 2 ones = ■ tens 12 ones
6. 2 tens 3 ones = 1 ten ■ ones
7. 7 tens 8 ones = ■ tens 18 ones
8. 4 tens 6 ones = 3 tens ■ ones
9. 5 tens 1 one = ■ tens 11 ones
10. 9 tens 8 ones = 8 tens ■ ones
11. 2 tens 0 ones = 1 ten ■ ones
12. 8 tens 4 ones = ■ tens 14 ones

A	B	C	D	E	F
4	2	18	6	16	13
G	**H**	**I**	**J**	**K**	**L**
10	3	15	5	17	7

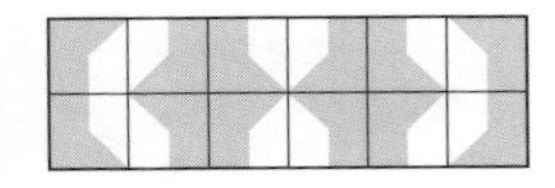

Objective: Regroup a 2-digit number as tens and ones.

Mending Differences

Find the difference.

1 $53 - 18$

2 $75 - 37$

3 $61 - 25$

4 $80 - 74$

5 $42 - 26$

6 $38 - 19$

7 $83 - 46$

8 $27 - 18$

9 What is 51 take away 36?

10 What is 60 take away 15?

11 What is sixty-one take away forty-three?

12 What is eighty-seven take away thirty-nine?

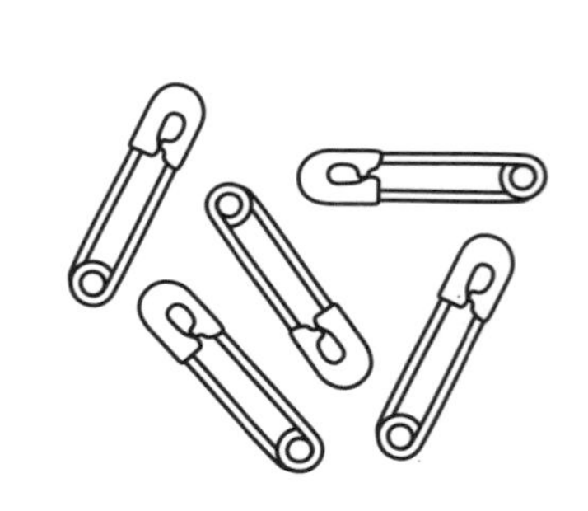

Answer Box

A	B	C	D	E	F
9	48	35	6	16	18
G	**H**	**I**	**J**	**K**	**L**
19	38	45	37	15	36

Objective: Find the difference of two 2-digit numbers with regrouping.

Let's Regroup!

Find the difference.

1 $67 - 19$

2 $56 - 27$

3 $72 - 45$

4 $96 - 58$

5 $43 - 36$

6 $70 - 49$

7 $82 - 45$

8 $60 - 29$

9 There are 44 apples on an apple tree. Al picks 27. How many apples are left on the tree?

10 There are 56 apples on an apple tree. Jo picks 28. How many apples are left on the tree?

11 Hiroshi and his friends pick 61 apples. They wash 43 of them. How many apples are left to wash?

12 There are 72 apple trees. 33 are picked. How many trees are left to be picked?

A	B	C	D	E	F
37	28	39	31	7	18
G	**H**	**I**	**J**	**K**	**L**
48	38	29	21	17	27

Objective: Find the difference of two 2-digit numbers with regrouping.

Round It Out!

Round each number to the nearest hundred. Then subtract to find the best estimate.

1
  252
− 186

2
  836
− 203

3
  631
− 315

4
  940
− 389

5
  871
− 117

6
  783
− 766

7
  551
− 441

8
  767
− 102

9
 1,000
− 105

10
  880
− 478

11
 1,199
− 210

12
 1,510
− 395

Answer Box

A	B	C	D	E	F
800	1,000	600	0	400	200
G	**H**	**I**	**J**	**K**	**L**
900	300	700	1,100	500	100

Objective: Find the estimated difference between two 3-digit numbers or between a 3-digit number and a 4-digit number.

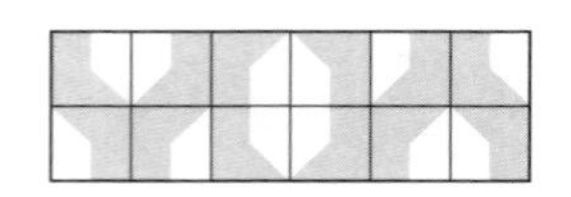

Nothing's Changing

Find the missing number.

1. 2 hundreds 1 ten = 1 hundred ▢ tens
2. 6 hundreds 4 tens = 5 hundreds ▢ tens
3. 8 hundreds 6 tens = 7 hundreds ▢ tens
4. 7 hundreds 3 tens = 6 hundreds ▢ tens
5. 3 hundreds 5 tens = ▢ hundreds 15 tens
6. 9 hundreds 7 tens = ▢ hundreds 17 tens
7. 6 hundreds 2 tens = 5 hundreds ▢ tens
8. 8 hundreds 5 tens = 7 hundreds ▢ tens
9. 3 hundreds 7 tens = 2 hundreds ▢ tens
10. 8 hundreds 4 tens = ▢ hundreds 14 tens
11. 4 hundreds 3 tens = ▢ hundreds 13 tens
12. 5 hundreds 9 tens = 4 hundreds ▢ tens

Answer Box

A	B	C	D	E	F
8	14	17	16	11	13
G	**H**	**I**	**J**	**K**	**L**
15	19	3	2	12	7

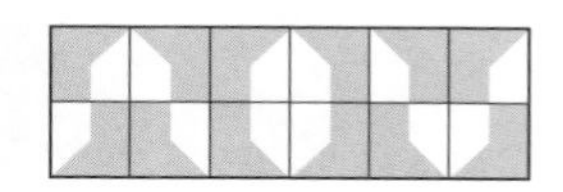

Objective: Regroup a 3-digit number as hundreds and tens.

Lost Money

Find the difference.

1. $\begin{array}{r} \$342 \\ -\ 128 \\ \hline \end{array}$

2. $\begin{array}{r} \$674 \\ -\ 338 \\ \hline \end{array}$

3. $\begin{array}{r} \$861 \\ -\ 527 \\ \hline \end{array}$

4. $\begin{array}{r} \$720 \\ -\ 619 \\ \hline \end{array}$

5. $\begin{array}{r} \$956 \\ -\ 438 \\ \hline \end{array}$

6. $\begin{array}{r} \$483 \\ -\ 264 \\ \hline \end{array}$

7. $\begin{array}{r} \$550 \\ -\ 321 \\ \hline \end{array}$

8. $\begin{array}{r} \$482 \\ -\ 139 \\ \hline \end{array}$

9. $\begin{array}{r} \$756 \\ -\ 228 \\ \hline \end{array}$

10. $\begin{array}{r} \$663 \\ -\ 439 \\ \hline \end{array}$

11. $\begin{array}{r} \$330 \\ -\ 209 \\ \hline \end{array}$

12. $\begin{array}{r} \$552 \\ -\ 206 \\ \hline \end{array}$

Answer Box

A	B	C	D	E	F
$214	$101	$518	$121	$219	$336
G	**H**	**I**	**J**	**K**	**L**
$229	$224	$528	$334	$343	$346

Objective: Find the difference of two 3-digit money amounts with regrouping tens as ones.

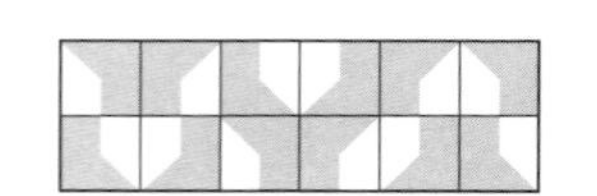

Giving It All Away

Sometimes you need to regroup hundreds as tens.

Find the difference.

1 428 − 293

2 749 − 362

3 941 − 470

4 522 − 341

5 856 − 594

6 873 − 683

7 721 − 150

8 460 − 180

9 643 − 281

10 727 − 492

11 551 − 270

12 839 − 352

Answer Box

A	B	C	D	E	F
281	262	235	571	280	487
G	**H**	**I**	**J**	**K**	**L**
471	362	135	181	190	387

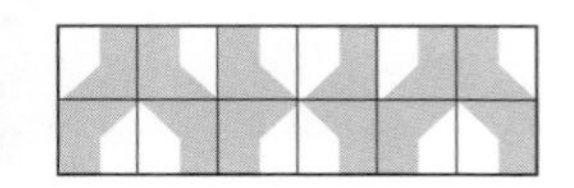

Objective: Find the difference of two 3-digit numbers with regrouping hundreds as tens.

Solve with Subtraction

Solve.

1 The ranch has 58 white horses and 27 brown horses.
How many more white horses than brown horses are there?

2 175 people rode at the ranch during June.
216 people rode during July.
How many more people rode during July than June?

3 The ranch used 157 bales of hay on Monday.
It used 173 bales on Tuesday.
How many more bales were used on Tuesday?

4 Juan worked 35 hours the first week.
He worked 42 hours the second week.
How many more hours did he work the second week?

5 The ranch has 42 mules.
It has 63 horses.
How many more horses are there?

6 44 people rode in a day.
26 rode in the afternoon.
How many people rode in the morning?

7 The library has 245 books about horses.

It has 316 books about horses or mules.

How many books are there about mules?

8 138 people rode in the Horse Show on Friday.

347 rode on Saturday.

How many more people rode on Saturday than on Friday?

9 There are 257 fence posts in the east field.

There are 341 posts in the north field.

How many more fence posts are in the north field?

10 There are 447 ranches in the country.

There are 229 farms in the country.

How many more ranches are there than farms?

11 34 children come to ride at 1:00 P.M.

17 children come at 3:00 P.M.

How many more children come at 1:00 P.M.?

12 The Silver Mountain trip is 88 miles.

The Owl River trip is 99 miles.

How many more miles is the Owl River trip?

Answer Box

A 18	B 41	C 218	D 71	E 16	F 84
G 209	H 11	I 7	J 31	K 17	L 21

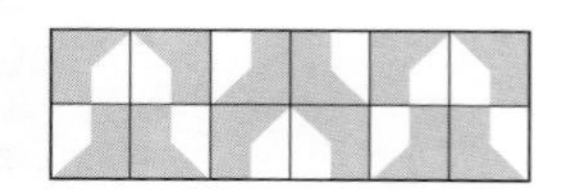

Objective: Solve a problem that involves subtraction.

Related Facts

Find the sum or difference.

1 37 + 72 = 109, so
109 − 72 = ■

2 341 + 528 = 869, so
869 − 528 = ■

3 91 − 36 = 55, so
55 + 36 = ■

4 481 − 263 = 218, so
218 + 263 = ■

5 60 + 39 = 99, so
99 − 39 = ■

6 72 + 85 = 157, so
157 − 85 = ■

7 53 + 82 = 135, so
135 − 82 = ■

8 90 − 45 = 45, so
45 + 45 = ■

9 73 − 28 = 45, so
45 + 28 = ■

10 621 + 181 = 802, so
802 − 181 = ■

11 431 − 207 = 224, so
224 + 207 = ■

12 528 + 403 = 931, so
931 − 403 = ■

Answer Box

A	B	C	D	E	F
481	621	72	341	60	73
G	**H**	**I**	**J**	**K**	**L**
53	37	528	90	91	431

Objective: Use a related addition or subtraction fact to find a sum or difference.

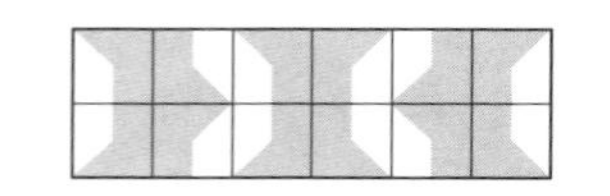

What Should We Do?

Find the sum or difference.

1 $170 - 65$

2 $57 - 32$

3 $341 + 218$

4 $86 + 75$

5 $72 - 48$

6 $47 - 36$

7 $63 - 58$

8 $47 + 36$

9 What number is twenty-eight more than six?

10 What is 260 take away 145?

11 What is the sum of 362 and 157?

12 How much larger than three hundred ninety-one is six hundred twenty-six?

Answer Box

A 24	B 34	C 25	D 11	E 161	F 115
G 559	H 519	I 83	J 235	K 5	L 105

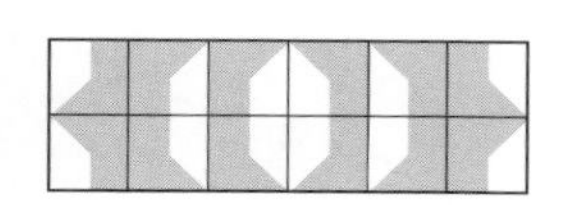

Objective: Find the sum or difference of 2- or 3-digit numbers.

Saving and Spending

Find the sum or difference.

1 $\begin{array}{r} \$43 \\ -\ 21 \\ \hline \end{array}$

2 $\begin{array}{r} \$354 \\ +\ 463 \\ \hline \end{array}$

3 $\begin{array}{r} \$155 \\ -\ 68 \\ \hline \end{array}$

4 $\begin{array}{r} \$78 \\ +\ 58 \\ \hline \end{array}$

5 $\begin{array}{r} \$229 \\ +\ 468 \\ \hline \end{array}$

6 $\begin{array}{r} \$429 \\ -\ 118 \\ \hline \end{array}$

7 $\begin{array}{r} \$968 \\ -\ 141 \\ \hline \end{array}$

8 $\begin{array}{r} \$339 \\ -\ 213 \\ \hline \end{array}$

9 Mark's family spends $195 for food. They spend $27 for gas. How much do they spend in all?

10 Marta's family put $465 in the bank. Sal's family put in $229. How much more did Marta's family put in?

11 Sue saved $18 in one month. She saved $13 the next month. How much did she save in all?

12 José had $149 in the bank. He took out $52. How much was left?

Answer Box

A	B	C	D	E	F
$97	$126	$87	$222	$236	$827
G	**H**	**I**	**J**	**K**	**L**
$817	$311	$31	$697	$136	$22

Objective: Find the sum or difference of 2- or 3-digit money amounts.

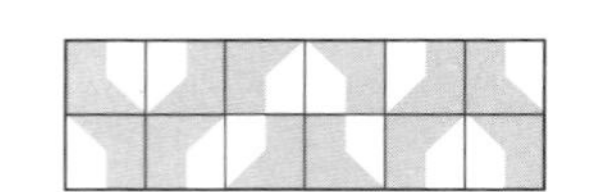